Obsolete
The Education Wake-Up Call

Sarah Kissane

"We taught them to follow the map.
Then changed the world beneath their feet."
— Sarah Kissane

Copyright © 2025 Sarah Kissane All rights reserved.

No part of this publication may be reproduced, stored in a retrieval system, or transmitted in any form or by any means — electronic, mechanical, photocopying, recording, or otherwise — without the prior written permission of the author, except in the case of brief quotations used in reviews or scholarly work.
This book is a work of nonfiction. Names and identifying details may have been changed to protect privacy.

Cover design by Sarah Kissane
Interior design and formatting by Sarah Kissane
ISBN: 978-1-7640829-0-7
First edition www.sarahkissane.com

Dedication

For my son, Maximilian.

I see you. I hear you. I believe in you.

The sky isn't the limit — it's just the beginning.

Table of Contents

Introduction – A World That's Already Changed 1

Chapter 1– A System Built for Yesterday 5

Chapter 2 – When Machines Learn Faster Than Our Kids Can 11

Chapter 3 – Education's Blindspot: The Human Advantage . 17

Chapter 4 – Your Child is Not a Robot 23

Chapter 5 – Reclaiming Wonder in a World That's Forgotten How to Play. It used to be simple. 29

Chapter 6 – The New Literacy: Teaching What Actually Matters in a World of AI 33

Chapter 7 – The Cost of Ignoring the Shift: What Happens If We Do Nothing 41

Chapter 8 – Where the Light Still Lives — In the Hearts and Minds of Our Teachers 45

    The Gap Between Knowing and Doing 46

    A Profession, Not a Sacrifice 47

    To the Ones Who Still Hold the Flame 48

Chapter 9 – Future Jobs: What Are We Really Preparing Them For? 51

    The Difference Between What's Useful and What's Just Tradition 53

    What We Can Do Right Now: Practical Steps to Prepare Our Kids 54

Careers That Sound Like Science Fiction—But Are Already Hiring 55

What We're Really Preparing Them For 56

Chapter 10: From Action to Vision — Rebuilding Education With Purpose 57

The Turning Point 57

Part I: What We Can Do — Today 57

Parents: Architects of the Everyday 57

Educators: Designers of Possibility 58

Communities: Co-Creators of Culture 59

Part II: What We Must Build — Together 60

A Day in a Future Worth Building 60

Curriculum That Breathes 60

Assessment That Elevates 61

Teachers as Trusted Professionals 61

Technology That Extends Humanity 61

Wellbeing as Infrastructure 61

Global Classrooms, Local Roots 62

Equity, Trust, and Shared Stewardship 62

What It All Adds Up To 62

Chapter 11: Final Thoughts – Choosing Courage 65

Introduction – A World That's Already Changed

Across the globe — in city apartments and rural towns, in private schools and public classrooms — something quietly devastating is happening to our children. It doesn't matter where they live, how much their education costs, or whether they wear uniforms or not. The truth is universal: we are preparing them for a world that no longer exists. School, as we know it, is failing them. And that failure is about to accelerate.

We are standing on the edge of the greatest technological shift in human history. Artificial Intelligence is no longer a prediction — it is our present. From the moment we wake to the moment we fall asleep, AI is shaping our choices, influencing our work, automating our systems, and changing the very definition of what it means to be productive, valuable, and employable.[1]

Yet our education system still trains children as though the world is static — as if the roles and rules of the 20th century are waiting for them when they graduate. This isn't just a lag. It's a catastrophic mismatch.

Our children are being taught to memorise, comply, and absorb — when what they will need most is to think critically, adapt quickly, and question boldly. They are graded on silence, stillness, and sameness — while the future rewards curiosity, collaboration, and creativity.

A classroom of desks lined in rows.

---

[1] **McKinsey & Company**, *The State of AI in 2023: Generative AI's Breakout Year* (2023).

Introduction – A World That's Already Changed

A bell dictating time.
A teacher as the holder of knowledge.
This isn't innovation. It's replication.

And in the age of AI, where machines can outperform humans in processing, calculation, and recall — the very things we still test for— this replication becomes dangerous. Because while we train our children to be perfect machines, the machines are becoming more perfect than we ever imagined. This book is a reckoning. It's for the parents who feel it in their gut but don't have the words. It's for the educators suffocating under standardised tests and outdated curriculums. It's for the students who already know — deep down — that the system doesn't see them, doesn't understand them, and isn't built for them. And it's for every one of us who believes that childhood should be more than preparation for an obsolete job market.

Whether your child lives in a high-rise or on a hilltop, whether they attend an elite school or one underfunded and overcrowded — the system is cracking beneath them. We can't afford to pretend it's fine. The evidence is everywhere: anxiety, disengagement, burnout, dropout rates, the rise of alternative education movements, and a tech revolution that's sprinting ahead while education crawls behind.[2]

Even without AI, the system was failing too many children. But now, the speed of technological change makes inaction a form of neglect. The OECD's international education data shows Australia is falling behind — particularly in equity, emotional wellbeing, and system adaptability, according to Education at a

---

[2] **World Economic Forum**, *The Future of Jobs Report* (2023).

Glance 2022: OECD Indicators (2022).[3]

We'll unpack education across time and place—from the rigid models inherited from the Industrial Revolution to the more progressive, child-centred approaches thriving in pockets of Europe today. In the Netherlands, for example, classrooms allow movement and agency.[4] Older students mentor the younger students. Lessons are experiential, and learning is collaborative, not competitive.

Contrast that with the silence of exam halls, the pressure of NAPLAN and SATs, the shrinking of arts and play to make room for literacy benchmarks.[5] Consider this: are children being raised as learners, or simply measured on how well they follow instructions?

This book will examine not just what schools teach — but what they ignore. Financial literacy, emotional intelligence, ethics in technology, conflict resolution, environmental consciousness, innovation, rest, how to think, how to question, how to lead, how to lose and how to try again. This is about real children in our schools—bright sparks whose potential fades before it ever gets the chance to shine; whose spirits are quietly dimmed before their talents even emerge. This book is not a complaint, it's a call. Because in the end, this isn't just about preparing our children for a better education — it's about preparing them for life. And life, as we now know, will look nothing like the past. If we want

---

[3] **OECD,** *Education at a Glance 2022: OECD Indicators* (2022).
[4] **HundrED,** *Innovations in Education: Netherlands Case Studies* (2022).
[5] **Australian Curriculum,** *Assessment and Reporting Authority (ACARA), NAPLAN Review Final Report* (2020).

our children to thrive, not just survive, we must build something better. Something honest. Responsive. Human. Let this be the beginning.

Chapter 1– A System Built for Yesterday

Step into almost any classroom today, and you'll witness a scene that would look familiar to someone from the 1950s — or even the 1850s. Rows of desks. A teacher at the front. A whiteboard or smartboard replacing the blackboard. Maybe laptops now instead of ink and slate. But the structure? Nearly identical.

The system may have added tech, but it hasn't evolved. And therein lies the problem. Today's children are being taught in a system built for a world that's already disappeared. Of course, education has existed for thousands of years — from the philosophical academies of Ancient Greece to the Confucian schools of China. But modern mass education — the system we see today — was largely shaped during the Industrial Revolution. It was never designed to foster innovation or creativity. It was built to meet the economic needs of that time: to produce obedient, punctual, and uniform workers.

It was a system of sorting — who would lead, who would follow, who would serve, and who would be discarded. Children were categorised by age, not by ability. They were expected to sit still, follow rules, respond to bells, and repeat back information. The primary goal was compliance and standardisation — not connection, self-discovery, or critical thought. This design was never about the child. It was about feeding the economy.

And yet, more than 150 years later, we still measure intelligence with multiple-choice tests. We still punish movement. We still ask children to memorise facts they could Google in seconds and regurgitate content they don't care about just to pass a test. We are asking them to be good at being schooled — not good at life.

Let's talk about how this system impacts children in the real world. Consider the six-year-old boy who can't sit still for more than ten minutes — not because he's disobedient, but because his brain is wired to learn through movement. Today, he's likely to be labelled with ADHD before anyone considers whether the classroom environment itself is the problem.

By the time he's eight, he believes he's "bad at school." By ten, he's disengaged. No one asked what lights him up. No one noticed that he can build anything with his hands or memorise the parts of a motorbike after one afternoon in the garage. The system only cared if he could sit, write, and listen — for hours on end.

And here's the truth we don't talk about: even as adults, many of us struggle with this model. We dread all-day meetings, lose focus in long lectures, and crave movement, variety, stimulation. Why do we expect more from six-year-olds than we can manage ourselves?

This story is heartbreakingly common.

Or the ten-year-old girl who excels in group projects but panics during timed exams. Who writes beautiful stories at home but freezes under the pressure of rigid literacy assessments. Who's told her future depends on her ability to recite persuasive writing techniques instead of writing from her heart. What are we teaching her? That the system's version of success is more important than her voice. Globally, we're seeing a growing disconnect between what schools teach and what the world needs.

Obsolete

A 2023 World Economic Forum report identified the top skills needed in the future workforce: analytical thinking, creativity, emotional intelligence, resilience, flexibility, and leadership.[6] But how many school systems are actually nurturing these? Instead, students are still rewarded for obedience, memorisation, and test performance. And let's talk about testing because it dominates modern schooling. NAPLAN (Australia), SATs and GCSEs (UK), ATAR scores (Australia), and the SAT and ACT in the United States. Entire school years are shaped around preparing for standardised exams that reflect very little of what a child actually knows or can do.

Standardised testing is not neutral. It benefits a narrow group of learners — and marginalises those who think differently, learn differently, or come from different backgrounds. It suppresses the very creativity machines can't replicate.

Some of the world's most impactful thinkers and leaders — from Albert Einstein[7] to Oprah Winfrey[8] — didn't thrive because of traditional schooling, but often in spite of it. They succeeded because of imagination, vision, and emotional depth — qualities that can't be bubbled in on a scan sheet. No one asks a child during these tests if they're hungry, anxious, neurodiverse, grieving, or carrying trauma. The system doesn't care. It just marks. But it's not just about what's broken. It's also about what's missing.

---

[6] **World Economic Forum**, *The Future of Jobs Report* (2023).
[7] Referenced in numerous biographies; most notably discussed in Walter Isaacson, *Einstein: His Life and Universe* (2007).
[8] Referenced in Winfrey's official biography and multiple educational profiles detailing her early schooling experience.

We don't teach children how to collaborate in a world that demands teamwork. We don't teach financial literacy, even though many will graduate into a life of debt. We don't teach consent, negotiation, or digital safety in any meaningful way — despite the fact that children today are growing up online.

We rarely allow space for failure, rest, or creative risk-taking. Yet those are the exact traits the future will demand. Failure, in particular, is where some of our greatest lessons are learned — it builds resilience, sharpens insight, and fuels innovation. But schools rarely allow space for it. We're still rewarding what machines already do best: repetition, memorisation, and rule-following. And that's a losing game. AI now outperforms us in every task that is linear, predictable, or fact-based. But it cannot imagine. It cannot connect. It cannot feel. The human edge lies in our ability to create, relate, and adapt. So why aren't we teaching that?

The hard truth is this: our education system isn't failing because it's broken. It's failing because it's doing exactly what it was designed to do — just in a world where its original purpose is now irrelevant.

The question we must ask ourselves isn't "how do we fix the system?" It's "why are we still using this system at all?"

Education should be about nurturing potential — not managing behaviour. It should be about lighting fires, not filling vessels. It should evolve as the world evolves. And yet it hasn't.

We wouldn't give our children a mobile phone from 1995 and call it cutting edge. We wouldn't teach them to drive using a horse and carriage. So why do we still use a 19th-century model to prepare them for a 21st-century world?

The stakes are too high now. This isn't just about frustration. It's about justice. It's about equity. It's about building a system that recognises every child — not just the ones who can sit still, follow rules, and tick boxes. It's about rewriting the purpose of education entirely.

Every human being needs a sense of purpose — something to work toward, something that matters. Education should be where that sense of purpose is discovered, nurtured, and celebrated. Because if we don't, we're not just failing a few kids. We're failing all of them.

Chapter 2 – When Machines Learn Faster Than Our Kids Can

We used to talk about the future as if it were far away — something our children would encounter one day, decades from now. We believed we had time to prepare them. Time to think. Time to catch up. But that future is already here.

Artificial Intelligence is not a distant concept. It's not science fiction. It's now. It's in our homes, our pockets, our workplaces, our healthcare systems, our shopping carts — and even in our children's classrooms, whether we acknowledge it or not. And yet, many people still don't fully understand what AI is. For some, it's a buzzword they hear on the news but can't quite explain. For others, it's something futuristic — like something out of The Matrix, Ex Machina, or *Her* — fascinating, but far removed from everyday life.

But AI isn't some far-off dystopia or cinematic fantasy. It's already here, woven quietly into our routines.

At its core, AI is simply technology that can learn, adapt, and make decisions — often mimicking human tasks. It's the voice that answers your questions on your phone. The algorithm that curates your social media feed. The software that scans your résumé or powers your smart home. It's already shaping how we live — and shaping the world our children will grow into.

AI learns fast, relentlessly fast. Faster than any human. Faster than any curriculum can keep up with. In fact, ChatGPT went from passing high school-level tests to achieving near top-percentile scores on bar exams and medical assessments — all within months of its release.[9] In a single week, AI models can digest and analyse more information than a student might encounter in an entire year of schooling. The pace is not just rapid — it's exponential. For many, this reality is still completely invisible.

The truth is, we are not just witnessing a technology shift — we are living through a **memorisation shift**. It's one of the most profound transformations of the modern era — bigger than the internet, bigger than smartphones, possibly even bigger than the printing press. In 2022, the world's most advanced AI models were doubling in power every few months — not years.[10] That means what seemed impossible at the start of the school year may be entirely outdated by the end of it. This is not evolution. This is revolution — and most people don't even see it happening. AI is evolving exponentially, faster than anyone can truly grasp. The job market our children are being prepared for may not exist in five or even two years—and many don't see it coming. Roles are being automated at an unprecedented rate. Industries are being disrupted before our eyes. What we once called "secure careers" are evaporating or being redefined altogether — not slowly, but rapidly, sometimes overnight.

So why is education still pretending it has time?

Let's look at the numbers. A 2023 McKinsey report estimated that 30% of hours worked across the U.S. economy could be automated by 2030.[11] Goldman Sachs predicts AI could impact

---

[9] **OpenAI**, *GPT-4 Technical Report* (2023).
[10] **Epoch AI**, *AI and Compute Trends Report* (2022).
[11] **McKinsey & Company**, *The Economic Potential of Generative AI: The Next*

Obsolete

300 million jobs globally.[12] Many of those jobs are white-collar. The kind our children are told to work hard for. ChatGPT and other large language models can already write essays, answer test questions, create code, solve equations, and generate content — in many cases, better than students, and sometimes better than their teachers. And that's just the beginning.

We now have AI-powered apps that can create entire marketing campaigns, edit videos, produce original songs, design logos, write business proposals, and more — often in minutes. If you can think of it, there's likely already an AI tool that can do it. We're not heading toward that future — we're already living in it. Meanwhile, schools are still focused on handwriting assessments. On memorising definitions. On teaching children how to follow the same steps as everyone else. We are preparing kids to compete with machines — in areas where machines are already winning.

This is not just short-sighted. It's negligent.

There are still things AI cannot do: it can't feel. It can't connect. It can't dream. It can't imagine something truly new. The human advantage lies in our ability to synthesise emotion, intuition, experience, and context. To collaborate. To empathise. To wonder. To create from nothing.

So why isn't our education system doubling down on that? Why are we still testing children on advanced spelling when autocorrect exists? Why are we forcing them to memorise

---

Productivity Frontier (2023).
[12] **Goldman Sachs**, *The Potentially Large Effects of Artificial Intelligence on Economic Growth* (2023).

formulas they'll never use, instead of teaching them how to think through real-world problems, navigate uncertainty, and challenge the status quo?

Here's the scariest part: this gap — between what AI can do and what schools still teach — is widening every day. Our children aren't just falling behind. They're being left behind.
Education systems around the world are too slow to respond. Governments — the ones who dictate education policy and approve curriculum — operate in bureaucratic cycles that move at a snail's pace. Policies take years. Curriculums take decades to shift. Meanwhile, technology is leaping forward by the month, often without their understanding or acknowledgment of its true impact.

Most children are learning to use computers, access AI tools, and navigate the internet. Some students are even using AI goggles or immersive learning headsets that place them lightyears ahead of the people teaching them. They know their way around digital platforms better than many of their teachers — and often, their own parents.

The result is a dangerous lag between what children are taught and what the world urgently demands they know. This creates what I call the adaptation gap — a chasm between what children are taught and what they'll actually need. And unless we build a bridge — fast — we risk raising a generation of young people who are deeply unprepared for the world they're inheriting.

They will be anxious, not because they're weak, but because we didn't give them the tools to thrive.

Let's bring this down to the personal level.

Imagine a teenager sitting in class, bored out of their mind, being taught how to write a five-paragraph essay. At home, that same teen is using AI to write code for a game they're building. Running a YouTube channel. Collaborating on Discord. Learning real-time editing, branding, marketing, and storytelling — skills that are relevant, dynamic, and increasingly valuable in the real world. Which one do we reward?

Too often, we dismiss what kids learn outside of school as play. But what if that is the real learning — and school is the outdated part? The system punishes the child who doesn't fit the mold. But maybe the mold is the problem.

This isn't about letting go of academic rigour. It's about redefining it. It's about asking deeper questions: what does it mean to be educated in the age of AI? What does it mean to be intelligent? What will it mean to be human in a world where machines can mimic almost everything — except soul?

We should be teaching:

- Ethics in technology — because our kids will help decide how AI is used.
- Creativity and design thinking — because original thought is the last frontier.
- Emotional intelligence — because relationships will remain irreplaceable.
- Adaptability and resilience — because change is now constant.
- Purpose — because without meaning, none of it matters.

Parents may ask: "But what can I do? I'm not a tech expert." You

don't have to be. You just have to be awake. Ask better questions. Advocate. Challenge your school. Support learning that feels alive. Create space at home for curiosity, creativity, and critical thought. Celebrate effort over outcome. Embrace mistakes. Talk about AI openly. Don't fear the future — prepare for it.
This chapter isn't about panic. It's about power. Because the sooner we acknowledge the truth — that machines are learning faster than our kids can — the sooner we can refocus education on what truly matters.

Not speed. Not perfection. Not memorisation.

But connection. Imagination. Humanity.

Chapter 3 – Education's Blindspot: The Human Advantage

Walk into a classroom almost anywhere in the world, and it echoes what we've already seen: children asked to sit still, be quiet, and focus on the same task at the same time in the same way. It's a pattern we've explored — one that prioritises compliance over curiosity, uniformity over uniqueness — and it still lingers as the root of education's greatest blind spot.

These aren't fringe traits. They are our evolutionary edge. In a world run by code, it is our feelings, our questions, and our art that still hold power. Empathy, creativity, and intuition aren't extras — they're essential. They're not peripheral; they're the future. Yet we continue to sideline them, as if leadership, compassion and imagination are optional. They are not luxuries or nice-to-haves. They are our species' edge. They fuel art, leadership, compassion, innovation — and still, we treat them as distractions rather than foundations.

The truth is, we've built an education system that overlooks the deepest part of what makes us human. We prioritise speed, accuracy, standardisation — machine traits — in a time when machines are doing those things better than we ever could. But what can't be replicated by AI? What can't be downloaded or automated or faked?

The human advantage.

It's not a new idea, but it's one we've forgotten. Einstein, who famously struggled in traditional schooling, once said:

"Imagination is more important than knowledge."[13] And he wasn't alone in that thinking. Pablo Picasso said, "Every child is an artist. The problem is how to remain an artist once we grow up."[14] These weren't just poetic musings. They were warnings. Warnings that creativity and individuality are often casualties of an education system more focused on testing than nurturing. And as we grow older, the damage deepens — we're taught to stop being silly, to stop asking "what if," to trade play for performance. We learn to be sensible, serious, and safe. We learn to think inside the box. Slowly, we begin to suppress the very instincts that allow us to create, innovate, and thrive.

Let's take a look at how that plays out. A child who doodles on the edge of their maths book isn't praised for imagination — they're told to "focus." A student who questions the logic of a history lesson isn't rewarded for critical thinking — they're marked as disruptive. A girl who writes poetry about grief, or a boy who builds entire cities in Minecraft, is rarely told, "This is brilliance. This is how your mind works." Instead, we chase scores. Rank them. Compare them. Reduce their uniqueness into band bubbles and baselines. But intelligence is not one-size-fits-all. And brilliance doesn't always wear a uniform.

Let's talk about Maya Angelou. She was mute for nearly five years after a traumatic event in childhood — and yet went on to become one of the most powerful voices in literature.[15] She turned silence into rhythm, pain into prose, humanity into

---

[13] **Albert Einstein**, widely attributed quotation in various writings.
[14] **Pablo Picasso**, frequently cited in educational psychology and creativity literature.
[15] **Maya Angelou**, *I Know Why the Caged Bird Sings* (1969).

poetry. Would the system have seen her as gifted? Frida Kahlo, after enduring physical trauma and isolation, used art as expression and defiance.[16] Her paintings weren't just beautiful — they were truth. Raw, emotive, disruptive. These are not anomalies. They are reminders: that pain, emotion, creativity, and reflection - the very things we don't measure in schools - can be the source of genius.

Today in classrooms across the globe, countless children are being misunderstood and overlooked simply because their minds work differently. A child with ADHD may be labelled disruptive—yet they're often deeply intuitive, brilliantly quick-thinking, and perfectly wired for active, movement-based learning. Neurodivergent students, who effortlessly recognise patterns invisible to others, find themselves penalised because their answers don't follow traditional methods or prescribed steps. Even daydreaming, the very birthplace of creativity and innovation—is frequently dismissed as distraction.

Our education system must recognise and nurture these differences — not silence them. We are teaching kids that staying inside the lines matters more than the ideas they want to express. Because when children are given creative freedom, they don't just express — they problem-solve. They innovate. They build resilience by experimenting, failing, adapting. They grow in confidence and curiosity. A blank canvas teaches them that their voice matters, their ideas belong, and the world is wide open to their imagination. That's not just how we nurture artists — it's how we raise future leaders, inventors, and changemakers. Meanwhile, AI can now mimic almost every task we once

---

[16] **Hayden Herrera**, *Frida: A Biography of Frida Kahlo* (1983).

considered uniquely human — writing, composing, problem-solving. But it still can't do this:

Feel the ache of loss in a poem. Draw from a well of lived experience. Cry while listening to music. Laugh at something unexpected. Comfort a friend.

These are not trivial things. They are the essence of humanity. The question is no longer "How do we help kids keep up with AI?" It's "How do we help them lean into the very things AI can't touch?" The real risk isn't just falling behind — it's pointing kids in the wrong direction entirely. AI isn't creeping in slowly — it's rewriting the rules of how we work, learn, and live.

If schools don't adapt now, we're not simply underpreparing students — we're underpreparing them for a future that's already arrived. Real-world employers are already seeing the cracks. A 2023 Deloitte report showed that empathy, collaboration, and adaptability are now considered more important than technical knowledge in many industries.[17] Google's own hiring studies have found that the most successful employees are not coders — they are communicators, critical thinkers, and team players.[18] And yet, schools still prioritise the opposite. We say we teach teamwork — and to some extent, we do. Children are encouraged to collaborate, participate in group projects, and be part of a team. But there's a quiet contradiction baked in. While we promote working together, we still rank and compare them individually. We push for unity, but reward

---

[17] **Deloitte**, *2023 Global Human Capital Trends Report* (2023).
[18] **Google / Project Oxygen**, findings widely referenced in organisational behaviour and hiring research.

competition. The message becomes confused: be a team player — but be the best. Stand out — but don't disrupt. We don't nurture emotional awareness — we label it sensitivity. We don't reward big-picture thinking. We tell them, "Stick to the rubric." And the heartbreak is this: so many children are already showing us who they are — and we're not listening. A boy who sits in silence — until music class, where his voice moves the room. A girl who struggles to spell — but spins entire worlds in her mind. A boy who fails every maths test — but can build a go-kart without a blueprint. These are not failures. They are gifts, and we're failing to see them.
So what do we do? We start valuing the human advantage again. We fight for it. We demand an education system that teaches:

Empathy — not as a buzzword, but as a fundamental life skill. It's the foundation of connection.
When we teach children to understand another person's feelings or perspective, we shape their ability to lead with compassion, resolve conflict peacefully, and work meaningfully with others. It reduces bullying, strengthens communities, and helps students become not just good learners, but good people. Empathy is what transforms classrooms into safe spaces — and societies into kinder ones.

Creativity — essential, not extra. It fuels innovation, helps us solve problems in unexpected ways, and gives rise to new industries, art forms, and ideas. It's how we build what's never existed before. When we nurture creativity, we give children the power to shape — not just survive — the future.

Curiosity — the root of learning, and the spark that dares us to dream. It fuels questions that change the world. It pushes boundaries, opens doors, and leads to unexpected discoveries. In every great breakthrough — scientific, artistic, or social — curiosity was the seed. When we encourage children to ask 'why' and 'what if,' we don't just educate them. We empower them to reimagine the world.

Intuition — true intelligence. It's not guesswork. It's the brain's ability to process complex, layered information quickly and insightfully. Intuition guides great decisions, sharpens judgment, and bridges logic with experience. It's how we "just know" — and in a rapidly changing world, that internal compass is invaluable.

Emotional fluency — a strength, not a weakness. To understand, name, and regulate one's emotions is a superpower. It leads to better mental health, stronger communication, and more authentic relationships. Emotional fluency helps children navigate conflict, build trust, and stay grounded when life gets messy — skills more important now than ever. Because this is what will future-proof our children.

Not more pressure. Not more exams. Not more conformity. *But more humanity.*

Chapter 4 – Your Child is Not a Robot

Children are not machines. And yet, somewhere along the way, we started treating them like they were. We optimise them. Monitor them. Schedule them to the minute. Feed them data. Track their performance. Measure their "output." Reward efficiency. Penalise pauses. Too often, we skip past how they feel and jump straight to how they score. We push them into structure before we've even taught them how to regulate their breath. Recent research reveals a troubling trend: diagnoses of anxiety, depression, and ADHD in children have sharply risen in the last decade. In Australia alone, almost 1 in 7 children aged 4–17 experience a mental health disorder each year, with ADHD and anxiety leading the list. Globally, the World Health Organization reports a 25% increase in anxiety and depression since 2020[19] — and children are not exempt. These aren't isolated cases — they're signals. Signals that our systems are pushing children beyond their emotional bandwidth. We act surprised when they break down, melt down, lash out or shut off. But it's not surprising. It's the system we've built. It's as if we've mistaken childhood for a coding problem to solve, a system to perfect, an operating model to streamline. But here's the truth: Your child is not a robot. They are not meant to perform on command. To power through without breaks. To absorb endlessly without space to reflect. They are not wired for relentless input, output, achievement. They are wired for movement, connection, emotion, expression and play. They are built for life. Not programming.

We are raising children in a world that celebrates hustle, speed,

---

[19] **World Health Organization**, *Global Burden of Disease: Mental Health Statistics 2023, Geneva*, 2023.

and productivity. It filters down from workplaces to classrooms to homes. We schedule every moment: sports, homework, tutoring, music lessons, language classes, exam prep, screen time rules, sleep trackers. We even count their steps — as if productivity can be measured by how many laps they run before recess. As if health, joy, and growth can be boxed into metrics.

And then we wonder why they're anxious. Why they can't sleep. Why they feel lost. Why their sense of self is so fragile it shatters under pressure. Because they're not being raised as people. They're being raised as products — confined within boundaries and rules from the moment they wake.

Their creativity is dulled at every corner, structured out of them through rigid routines and over-scheduling. Even play is now often organised, timed, and adult-led. Modern children are so accustomed to being directed that many have lost the simple, profound ability to use their imagination freely — to climb trees, build forts from sticks, or get lost in the joy of mud. As a parent, how often do you hear, "I'm bored," simply because they believe they have nothing to do unless it's been handed to them?

The world is out there — full of wonder, nature, stories, and possibility — but they are stuck in a system that doesn't let them think like that.

They're losing the capacity to just be.
To roam.
To play without rules.
To explore without limits.
And in that loss, something deeply human is being stripped away.

Obsolete

Let's go deeper.

Robots don't need to process grief. Children do.
Robots don't need to understand belonging. Children do.
Robots don't cry when they feel lonely, or ache when they fail, or spiral when they feel unseen.

But our education systems often behave as though these things are distractions — inconveniences to the greater goal of academic success.

So when a child has a panic attack in the middle of a test, we say, "They're not resilient enough." When a teenager withdraws emotionally because they can't keep up, we call it disengagement. But what if it's not the child who's malfunctioning? What if it's the system?

Let's sit with that.

No one expects a robot to be creative. Or to question the rules. Or to ask "what if?" or "why not?" But that's exactly what we should expect from our children. The reality is that many of the most remarkable, impactful individuals in history were brilliant precisely because they refused to become robots. Steve Jobs didn't follow the script. He dropped out, explored his passions, and trusted his instincts, revolutionising how we communicate, work, and create.[20] Richard Branson struggled with traditional schooling due to dyslexia—yet went on to redefine entrepreneurship through innovation, adventure, and

---

[20] **Walter Isaacson**, *Steve Jobs* (2011).

imagination.[21] Agatha Christie, one of the most influential novelists in literary history, was described by teachers as slow and struggled with spelling—but that didn't stop her from captivating millions worldwide.[22] Walt Disney was fired from a newspaper job for lacking imagination, only to redefine entertainment and storytelling forever.[23] None of these individuals were standard or typical. They changed the world because they embraced their uniqueness rather than suppressing it.

Yet today, our education system still funnels children towards Conformity. We rank them. Compare them. Label them as "gifted" or "below average." We hand them numerical identities before they even discover who they are inside. We praise speed and standardisation — and overlook the quiet dreamers and creative explorers.

But the future is built not by the obedient, but by the imaginative. By those who drift off into their thoughts during lessons because they are envisioning new possibilities. We instruct kids to sit still when they were born to move. We command them to memorise when their nature is to inquire. We reward conformity and penalise bold thinking that spills beyond the lines.

Let's be clear: this isn't about lowering expectations. It's about raising them — intelligently. Suppressing creativity is not just inefficient; it's detrimental. Ignoring emotional well-being isn't

---

[21] Frequently referenced anecdote, earliest attribution in various biographies and media history articles.

[22] **Laura Thompson**, *Agatha Christie: An English Mystery* (2007).

[23] Frequently referenced anecdote, cited in multiple biographies and Disney archives.

just irresponsible; it's costly. Convincing children their value depends on their productivity isn't just harmful; it's dangerous. We owe children more than love — we owe them the space to be fully human. Humanity is messy, emotional, unpredictable — and filled with brilliance and mistakes. If we treat our children like machines, we risk eroding the unique gifts they possess and burning out their innate potential.

Your child is not a robot. They don't need an upgrade. They need recognition. They need to be seen, valued, and deeply heard. That fundamental acknowledgment — knowing someone truly sees and appreciates them — is vital for confidence, resilience, and self-belief. They deserve nourishment, not optimisation. Encouragement, not calibration. Love, not measurement.

Chapter 5 – Reclaiming Wonder in a World That's Forgotten How to Play. It used to be simple.

You'd open the door and the kids would disappear — not onto screens, but into the backyard. Into trees and dirt and stick-swords. Riding bikes with no helmets, scraped knees that didn't need band-aids unless they were bleeding badly — and even then, only if they were dragged back inside by a concerned sibling. Play is not a break from learning. It is learning. It's how children test ideas, grow emotionally, and imagine new worlds.

That kind of play — spontaneous, free, wildly imaginative — is disappearing. And we're not just losing childhood. We're losing the fire that fuels everything else.

Let's be clear: play is not a luxury. It is not an add-on. It is not optional. Play is how children learn. It's how they process emotion, build problem-solving skills, develop empathy, take risks, and explore who they are. In countries like Finland and Estonia, formal academics are intentionally delayed until the age of seven, with a strong emphasis placed on learning through play.[24] These systems recognise that social, emotional, and imaginative development are not only foundational — they're essential to long-term academic and life success.

We see echoes of this philosophy closer to home as well. In Australia, many Steiner schools embrace similar values — prioritising creativity, emotional growth, movement, and imagination in the early years, rather than rushing children into

---

[24] **OECD**, *Starting Strong V: Transitions from Early Childhood Education to Primary Education* (2017).

formal academics.[25] But despite these examples, most mainstream systems don't treat it that way. Schools cut recess to fit in more literacy. Parents are told to prepare their toddlers for "school readiness" — a term that often means conditioning children to sit still, follow instructions, and perform rather than wonder. In many classrooms, snacks are restricted or skipped altogether, even though we know children can't function properly when they're hungry. What we call readiness is really just early initiation into a system of conformity — one that prioritises compliance over curiosity, and outcomes over imagination.

Play is scheduled, supervised, controlled — if it exists at all. And it's not just sad — it's alarming. We are stripping away the very foundation of childhood development in the name of academic performance and measurable results. It's a slow erosion of curiosity, joy, and emotional resilience — and the cost is showing up in our kids' anxiety, disconnection, and burnout. When we silence play, we're not just quieting fun — we're cutting off the lifeblood of learning, self-expression, and mental health. That's not just damaging. It's devastating.

In the early years, unstructured play literally builds the brain. Numerous studies in developmental neuroscience show that imaginative play activates and strengthens neural pathways associated with executive function — the core set of mental skills that include working memory, cognitive flexibility, and self-regulation. According to the Harvard Center on the Developing Child, these skills are critical for setting and achieving goals,

---

[25] **Steiner Education Australia**, *Educational Foundations and Learning Philosophy*, www.steinereducation.edu.au.

making decisions, and managing social interactions. A study published in the journal Pediatrics[26] found that preschoolers who engaged in more unstructured play demonstrated significantly better self-regulation, focus, and emotional control — key predictors of academic and life success. It also boosts creativity, resilience, and confidence. It helps children make sense of their world, their relationships, and their feelings. So when we take it away, we're quietly setting them up to fall behind in the things that matter most. That's not a flaw in them. That's a reflection of the environment we've built. We've over-scheduled their time. Replaced "go outside and explore" with "complete your extra worksheets." Traded imagination for outcomes. Swapped unstructured afternoons with rigid routines and endless reminders to finish homework — which schools often insist on — even when it means giving up the last precious moments of play. We are unintentionally teaching children that productivity is more valuable than joy, that being busy is better than being curious. But it's not too late.

In fact, the fix is beautifully simple: let them play. Let them get messy. Let them be loud. Let them climb and build and pretend. Let them be bored long enough to create something new. Don't fill every gap. Don't rush in with solutions. Don't correct the "wrong" way to play. Give them space. If you live in a small apartment, start with a box and some markers. Let them turn it into a spaceship. If you live in a city, visit a park with trees. Let them touch bark, watch ants, imagine fairy kingdoms under leaves.

---

[26] **Pediatrics**, *The Importance of Play in Promoting Healthy Child Development and Maintaining Strong Parent–Child Bonds* (2007).

If you're a teacher, create space for wonder. Build in ten minutes of "wild idea time." Let them question. Let them wander. Wonder is not bound by location. It's bound by permission. So give it. Because the world they're heading into will need more than knowledge. In fact, knowledge — once king — is rapidly being dethroned. With AI capable of retrieving, processing, and applying information faster and more accurately than any human, memorised knowledge is becoming increasingly obsolete. What remains irreplaceable is imagination. Courage. Collaboration. The kind of thinking that can't be taught — only uncovered through experience. And play is the birthplace of it all. So, let's bring it back. Not just for our children. But for all of mankind.

Chapter 6 – The New Literacy: Teaching What Actually Matters in a World of AI

We've reached a pivotal point in human history — one that demands a radical rethinking of what it means to be educated. Education once meant efficiency. But in a world that now values adaptability, imagination, and collaboration, that definition no longer fits. The disconnect between school and the real world is not theoretical — it's visible in every classroom, every day. We are still teaching for a world that has already changed. And children feel the strain of that misalignment, even if they don't yet have words for it.

The world our children are being educated for no longer exists — as I have pointed out previously, again and again, because this isn't a passing observation. It's a fundamental truth we can no longer ignore. The systems that shaped past generations were designed for a world of industrial jobs, linear progression, and predictable careers. Schools were built to create order, conformity, and proficiency. The goal was clear: learn the information, follow the rules, get the grade, get the job. But that model is now obsolete.

We are living through a seismic shift — a moment in history where artificial intelligence is not just matching human capability but redefining what it means to be exceptional. AI can now write with eloquence, code with precision, and analyse data at speeds no human could rival. It can diagnose illness, generate art, build websites, compose music, and hold conversations that feel almost therapeutic. It crafts marketing strategies, designs logos, and launches new business ideas before most leadership teams have even entered the boardroom. It can mimic human voices, faces, and gestures so convincingly that reality itself begins to blur. What once set us apart is no longer ours alone — and the implications are profound. So if AI can do all this — and more — what do we actually need our children to learn? What becomes valuable in a world where knowledge is cheap and automation is everywhere? The answer isn't found in more memorisation, stricter testing, or longer homework hours. The answer lies in what makes us beautifully and unmistakably human.

We need a new literacy. One that values:

Emotional intelligence over rote knowledge — the kind learned purely by repetition, without depth or real-world understanding. Rote learning may help pass a test, but it rarely prepares a child to navigate uncertainty or solve problems creatively.

Creativity over repetition — not just creating art, but thinking divergently, solving problems in new ways, and imagining what's possible beyond the obvious. Repetition may build routine, but creativity builds innovation.

Critical thinking over blind acceptance — the skill to analyse, evaluate, and challenge ideas. In a world flooded with information, it's no longer enough to know — kids must learn how to discern.

Collaboration over competition — not just working together, but co-creating, compromising, and listening. In a globally connected world, being able to work across perspectives is more powerful than winning alone.

Curiosity over compliance — asking 'why?' and 'what if?' instead of simply ticking boxes. Curiosity leads to discovery. Compliance alone leads to stagnation.

Purpose over performance — doing work that matters, not just work that's rewarded. Performance fades. Purpose fuels lifelong motivation and fulfilment.

Because while AI is growing exponentially, it still can't empathise with a friend. It can't feel awe in nature. It can't write with soul. It can't create with emotion. It can't lead with moral clarity. It can't build authentic trust through lived experience. It can't understand grief without data. It can't laugh at a moment that wasn't scripted. It can't sit in silence with someone and just know what they need. It can't dream — not just of what's possible, but of what's meaningful.

And these are the very things the future will need more of — not less. So what should we be teaching? Let's start with the most overlooked skill of all: curiosity. In a world where AI can find the answer, the power is in asking better questions. We need children who are questioners, not just answerers — kids who wonder and wander, who challenge, test, and reimagine the world with eyes wide open and minds unbound.

Then there's empathy and ethics — the ability to see the world through another person's eyes, to weigh consequences, and to choose kindness over convenience. These aren't soft skills — they are survival skills in a world where decisions will increasingly be made by algorithms. As machines grow in power, our moral compass must grow stronger, more deliberate, and more deeply rooted in our shared humanity. We teach formulas instead of friendships. Facts instead of self-awareness. And we wonder why so many graduates are unmotivated, emotionally fractured, and unsure of how to connect in the real world.

We don't have to look far to see this in action. In one longitudinal study, kindergarteners with higher emotional intelligence were significantly more likely to graduate high school, avoid legal trouble, and have full-time jobs two decades later.[27] Another found that EQ, more than IQ, predicted long-term career success,[28] especially in leadership, teaching, and caregiving professions. The science is clear — EQ doesn't just make good humans, it makes future-ready ones.

Digital and media literacy is no longer optional. Children must learn how to navigate algorithms, misinformation, online identity, digital footprints, and consent in a world where much of their lives are mediated by screens. They must understand the permanence of what they post, the risks of identity fraud, and that anything shared online could be accessible — forever. It's about knowing how the digital world works, how to stay safe, and how to think critically in a space designed to distract and exploit. Resilience and adaptability matter more than perfection. The pace of change is dizzying — our kids must know how to pivot, recover, and stay grounded in a world that won't stop shifting. And yet, while we scramble to teach children how to navigate the digital world, we often overlook the emotional world they're already struggling to survive in. Emotional intelligence is not a bonus skill — it's a critical survival tool in an

---

[27] **Jones, D. E., Greenberg, M., & Crowley, M.**, *Early Social-Emotional Functioning and Public Health: The Relationship Between Kindergarten Social Competence and Future Wellness*, American Journal of Public Health (2015).

[28] **Goleman, D.**, *Emotional Intelligence: Why It Can Matter More Than IQ* (1995).

age of rapid change, constant connectivity, and growing social complexity. When EQ is underdeveloped, we see it everywhere: rising anxiety, broken friendships, workplace conflict, burnout, and the inability to manage failure. The cost isn't theoretical — it's real, and it's rising. If we want children who can thrive — not just cope — we must teach them how to know themselves, soothe themselves, and connect meaningfully with others.

And perhaps most importantly, we must teach self-awareness and purpose. In a world where it's easy to feel small or irrelevant, children need to understand who they are, what matters to them, and how they can contribute meaningfully to the world around them. And in doing so, we also begin to return to a more cohesive, selfless society — one built on the mindset that it takes a village. That no one thrives alone. That our strength, both now and in the future, will come from connection, not competition.

Let's be real: most schools aren't teaching these things — not consistently, not with depth. And it's not because teachers don't care. Many care deeply. But they're exhausted. They lose the will to fight a battle they've already fought — and lost — time and time again. They've tried to drive innovation, to bring life into the classroom, to make meaningful change. But the system resists. The same system we've already acknowledged was built for a completely different era — one obsessed with performance metrics and standardised results, where the curriculum assumes every child learns the same thing in the same way, at the same pace, regardless of their individual abilities or needs. It often ignores whether students are truly learning, growing, or thriving.

We can't prepare children for the future with a curriculum built for the past. We need a shift. From subjects to skills. From instruction to exploration. From obedience to agency. And here's what's powerful: this doesn't need to start with policy. It can start at your dining table — a place where real conversations happen, where stories are shared, and where curiosity can be nurtured without a lesson plan. Ask them what they're curious about. Lean in. Let them lead. You may uncover interests and questions that surprise you — the kind that never show up on a school test. Talk to them about the big stuff: fairness, justice, purpose, fear, love, failure. These aren't just adult conversations — they're human ones. Help them name what they feel and think deeply about what matters.

Explore their dreams and future. Ask them what they want to do. Who they want to become. What the future looks like in their imagination. Many will say "a job" — often one they've seen adults do. But the reality is, that job might not even exist when they leave school.

Guide them toward possibility. Help them see beyond job titles. Show them how to build the core human skills that will serve them in any future. Even if AI takes their dream job, those skills — curiosity, empathy, critical thinking, adaptability — will be their superpower.

Teach life skills as future skills. Show them how to disagree respectfully. How to stand up for what's right. How to reflect when they're wrong, and how to get back up after failing. These aren't just soft skills. They're survival skills. They're future-proof. These are life skills — and they're future skills.

And if you're an educator, know this: you have more influence than you realise. You don't need permission to foster wonder. You don't need a new curriculum to build resilience. Every conversation, every encouragement, every moment you allow a child to think for themselves — that's revolutionary. You are not preparing them for exams. You are preparing them for life. The new literacy is about more than decoding words. It's about decoding the world. It's about teaching children how to think for themselves, question systems, recognise truth, communicate meaningfully, navigate change, and stay human in a world that often forgets to be. Because no matter how advanced AI becomes, it will never replace the power of a young person who knows who they are, thinks critically, and acts with heart. And that's what truly matters.

Chapter 7 – The Cost of Ignoring the Shift: What Happens If We Do Nothing

We need to say what no one wants to admit. If we keep going the way we are, our schools won't just feel outdated — they'll feel irrelevant. The world is moving forward, but we're standing still. And children feel the drag of that disconnect, every single day. We're going to lose our children — and with them, we risk losing the very essence of our humanity. Not physically. But emotionally, mentally, and spiritually — many of them are exhausted. The pace is unsustainable. The pressure, inescapable. Look around: we see the effects in classrooms across the globe — children struggling to stay engaged, to self-regulate, to feel safe being themselves.

Record numbers of teens are disengaging from education altogether. Support services are stretched thin. Waiting lists are long. Teachers are overwhelmed. And still, we ask them to do more with less. We see an epidemic of perfectionism and performance-based self-worth. A lack of desire to do hard work — not because children are lazy, but because the system has taught them that effort only matters if it leads to a grade or job. They've been conditioned to believe that working hard is transactional, a means to a narrow end, rather than something meaningful in itself. We see a growing disconnect between learning and relevance — where children fail to see how their education connects to the real world. A fear of failure so deeply ingrained that many avoid trying altogether, choosing to play it safe rather than take creative or intellectual risks — taught that mistakes are weaknesses, and that success only counts if it's perfect the first time.

And we wonder why. We wonder why our children are unmotivated. Why they're disconnected. Why they no longer believe in themselves. But we forget — they've been told to fit in, not stand out. They've been rewarded for obedience, not originality. They've been tested on memorisation, not meaning. And now, we're watching them quietly unravel. We're raising a generation trained to follow instructions, but unprepared to trust themselves. A generation that can recite facts, but doubts their own voice. A generation that is digitally fluent — but emotionally lost. If we do nothing — if we keep teaching the same way, assessing the same way, measuring worth the same way — we won't just raise underprepared workers. We'll raise undernourished humans. This isn't just being dramatic. It's happening.

Some children are being diagnosed not because something is wrong with them — but because something doesn't fit around them. We've built a system that expects stillness, sameness, and standardisation — and calls it disruption when a child resists.

The erosion of confidence starts early. When a child learns that raising their hand with the wrong answer earns them embarrassment instead of encouragement, they begin to withdraw. When we demand children stay neatly within the boundaries, we quietly erase the boldness that makes their creativity shine. When day after day they're compared to their peers instead of supported as individuals, their self-worth becomes transactional. And over time, the message is clear: play it safe. Follow the rules. Don't risk failing. Don't risk standing out.

We are teaching children to abandon their instincts. To mistrust their originality. And by doing so, we are quietly editing out the very qualities the future needs most.

The economic cost is real, too. If we continue to prepare children for yesterday's workforce, we'll create a society of misfits—not because our kids failed, but because we failed them. Reports from the World Economic Forum,[29] McKinsey,[30] and global think tanks all point to one truth: future jobs will demand problem-solving, digital fluency, human connection, and adaptability. Not standardised memory recall. The cost of doing nothing isn't just personal. It's economic. It's systemic. It's generational.

And what about the spark? You've seen it — the light in a child's eyes when they talk about something they love. Dinosaurs. Space. Music. Invention. LEGO creations stacked sky-high. Minecraft worlds they've built block by block. Questions about the world that seem too big for their little bodies to hold. What happens when we don't nurture that? It dims. Sometimes it goes out. When curiosity is ignored, when creativity is compressed, when learning becomes a box-ticking exercise, that spark — the thing that makes them them — flickers. And we lose more than we know.

So what happens if we don't change? We raise disconnected kids in a connected world. We raise anxious teens who measure their worth by performance. We raise graduates who don't know who they are, what they want, or how to handle rejection, failure, or change. Many are entering university not for a love of learning, but out of fear of falling behind. But will universities even have a place in the new world? What jobs are they going into debt for?

---

[29] **World Economic Forum**, *The Future of Jobs Report* (2023).
[30] **McKinsey & Company**, *The Economic Potential of Generative AI: The Next Productivity Frontier* (2023).

And will those jobs even exist by the time they graduate? And we wonder why we're losing them.
It's not too late — but it is urgent.

I'm not saying tear everything down. But I am saying: evolve. Be brave enough to question what we've accepted for too long. To reimagine education with children's whole selves in mind. The cost of inaction is not abstract. It's not theoretical. It's playing out in classrooms, in bedrooms, in counselling offices, in quiet moments where children say:

"I'm not good at anything." "I hate school." "I'm just not smart." "I don't belong."

We can't afford to keep hearing that. We have to listen. We have to act. Because the future is coming — and it deserves children who are ready to meet it, not just with qualifications, but with curiosity, character, and confidence.

Chapter 8 – Where the Light Still Lives — In the Hearts and Minds of Our Teachers

Most don't do it for applause. Many stay because they care too deeply to walk away. Teachers play a vital role in keeping the education system functioning. Across the world, many step into classrooms that are under-resourced and burdened by competing demands — and still, they do their best to show up for their students. For the connection. For the possibility of a spark. For the belief — however challenged — that education still has the power to make a difference. This chapter is for you.

For the teacher who gently wipes away a child's tears before teaching phonics. For those who buy glue sticks and snacks from their own salary. For those who stay up past midnight adjusting lesson plans to reach the child who's slipping away. For the ones who know a student's silence says more than a grade ever could. It's time we told the truth: the system hasn't survived in spite of teachers — it's survived because of them.

The weight they carry isn't just the job description that exhausts them. It's what's been added and subtracted over time. We've added admin, data-tracking, standardisation, behavioural documentation, performance monitoring, and parent diplomacy. We've subtracted autonomy, trust, rest, reflection, creativity, and joy — and it didn't happen overnight.

Over the past two decades, in response to global rankings, political scrutiny, and a push for performance-based accountability, education systems slowly traded trust for metrics. With every new policy, report, and reform agenda, something quietly disappeared from classrooms: the freedom to teach with humanity, with intuition, with fun and energy. And yet, teachers are expected to carry the emotional, intellectual, and logistical responsibilities of educating and nurturing entire cohorts of young people — often without adequate support, and with little room to say: this is too much. In Australia, attrition rates suggest up to 25% of early-career teachers leave within five years.[31] In the U.S., it ranges from 19% to 30%. These aren't numbers — they are people. Hearts that once burned bright for education now quietly fading out. Many don't leave because they've stopped loving teaching. They leave because teaching inside the current system requires a kind of self-erasure.

One teacher told me, "I used to be a person with purpose — now I feel like paperwork in motion." Another said, "I used to teach. Now I manage behaviours, meet KPIs, and hope there's time left to connect with a child." And it's not sustainable. Not for them. Not for any of us.

The Gap Between Knowing and Doing

Teachers are experts in their craft — but the system does not always reflect that truth. Policy decisions are often made far from the classroom, by those who have never stood in front of thirty students, let alone tried to reach them. Many know, both intuitively and professionally, what their students need. They read the room. They notice what's unsaid. But too often, they're boxed in by time constraints, standardised curriculums, and

---

[31] **Gallant, A., & Riley, P. (2014).** *Early career teacher attrition in Australia: Addressing the issue of early career burnout. Australian Journal of Teacher Education*, 39(4). In U.S. context: Gray, L., & Taie, S. (2015). *Public School Teacher Attrition and Mobility in the First Five Years* (NCES 2015-337). U.S. Department of Education, National Center for Education Statistics.

systems that value compliance over connection. The curriculum tells them what to teach. The assessments dictate how to measure it. The policy decides what matters. Their instincts — their humanity — are left on the cutting-room floor.

Still, they innovate. They adapt. They build connection where bureaucracy builds distance. They find space in the margins to remind children they matter. They are not overworked because they lack organisation. They are overworked because they are expected to carry more than any one human should — data, behaviours, parent communication, mental health crises, paperwork, performance targets — all while trying to light a spark that keeps learning alive.

This is the space between who they truly are, and who the system allows them to be. And the cost of that space is everything from morale to retention, from innovation to joy.

A Profession, Not a Sacrifice

We must stop romanticising their exhaustion. Burnout is not a badge of honour. Passion should not be punished. And martyrdom is not a sustainable model. Teaching is a profession — and should never demand self-sacrifice to prove its worth.

That means:

- Trusting teachers as designers of learning, not just deliverers of content.
- Paying them not just fairly — but generously.
- Protecting time for planning, collaboration, rest, and reflection.
- Involving them in policy, curriculum, and reform at every level.

- Offering support, not surveillance.
- Valuing their emotional labour as real, measurable, worthy work.

When teachers thrive, children do too. When teachers are inspired, empowered, and well — schools become places of life, not just survival. Wonder returns. Joy becomes visible. And the learning that unfolds carries light, not just obligation.

To the Ones Who Still Hold the Flame

Once, whilst teaching in a school, I stood at the back of a classroom and watched another teacher pause mid-lesson because a student was spiralling — not academically, but emotionally. She knelt beside him, whispered something I couldn't hear, and the entire room softened. She never raised her voice, never lost her rhythm. In that moment, I understood: educating is not just a job. It is an embrace without arms.

Even now, teachers are doing what the system has forgotten how to do:
- They're building belonging in classrooms that feel like home.
- They're teaching empathy between the maths lessons.
- They're lighting sparks that no test could ever measure.
- These are not small things. They are revolutionary acts — often unseen, but never unfelt.
- They are not just surviving a broken system. They are quietly redesigning it.
- They are not relics. They are architects.
- They are not burned out — they are burning bright despite it all.

And when the world finally catches up to what education could be — it will be because teachers held the flame long enough for the rest of us to see the light.

This is your chapter. Your courage. Your leadership. And your rightful place at the heart of what comes next.

Chapter 9 – Future Jobs: What Are We Really Preparing Them For?

The world of work is changing. Rapidly.

As we stand on the edge of a technological and societal revolution, it's becoming increasingly clear that the jobs our children will one day step into may bear little resemblance to those we see today. Roles that were once considered safe or even prestigious are being transformed, displaced, or replaced altogether. Simultaneously, entire industries are springing to life, built upon the foundations of AI, renewable energy, quantum computing, synthetic biology, space exploration, and other rapidly advancing fields.

The speed of this transformation is staggering. Automation, climate change, demographic shifts, and exponential technological growth are reshaping every sector — from healthcare to education, logistics to law. For parents, educators, and society at large, this demands a shift in how we think about education, career readiness, and what it truly means to prepare our children for the future.

What does this future look like? What roles will matter most? What knowledge and human qualities will remain relevant in an age of intelligent machines and planetary urgency?

This chapter dives deep into the critical jobs of tomorrow — and the mindset required to navigate this new world. It's not just a list — it's a call to evolve our thinking, our systems, and the way we raise and educate our kids.

According to the World Economic Forum's Future of Jobs report,[32] the workforce is undergoing rapid transformation — with key
industries rising, others disappearing, and the skill demands shifting faster than ever before. Report 2025 captures the swift, transformative shifts reshaping careers and opportunities.[33]

Future of Jobs Report 2025

Fastest growing and declining jobs by 2030

WORLD ECONOMIC FORUM

#	Top fastest growing jobs	Top fastest declining jobs
1	Big data specialists	Postal service clerks
2	FinTech engineers	Bank tellers and related clerks
3	AI and machine learning specialists	Data entry clerks
4	Software and applications developers	Cashiers and ticket clerks
5	Security management specialists	Administrative assistants and executive secretaries
6	Data warehousing specialists	Printing and related trades workers
7	Autonomous and electric vehicle specialists	Accounting, bookkeeping and payroll clerks
8	UI and UX designers	Material-recording and stock-keeping clerks
9	Light truck or delivery services drivers	Transportation attendants and conductors
10	Internet of things specialists	Door-to-door sales workers, news and street vendors, and related workers
11	Data analysts and scientists	Graphic designers
12	Environmental engineers	Claims adjusters, examiners and investigators
13	Information security analysts	Legal officials
14	DevOps engineers	Legal secretaries
15	Renewable energy engineers	Telemarketers

Note: The jobs that survey respondents report the highest and lowest net growth (%) by 2030.
Source: World Economic Forum. (2025). Future of Jobs Report 2025.

Are we preparing children for the future—or anchoring them to the past? Recent research by McKinsey & Company indicates that

---

[32] **World Economic Forum**, *The Future of Jobs Report* (2023).
[33] **World Economic Forum**, *Future of Jobs Report 2025* (2025).

by 2030, Europe and the United States could each require up to 12 million occupational transitions due to automation and the adoption of generative AI. Beyond this snapshot lies an urgent truth: the future isn't arriving—it's already here. Jobs once deemed secure and prestigious are evaporating like morning mist, replaced by roles we scarcely imagined could exist. If education remains anchored to yesterday's methods, we risk sending our children into tomorrow blindfolded.

Today's schools still emphasise memorisation, uniformity, and compliance — a practice starkly misaligned with the World Economic Forum's recent findings that nearly half of today's essential job skills will dramatically change by 2027. [34] Skills designed for factories, not futures. Yet emerging roles demand exactly what machines can't replicate: creativity, empathy, lateral thinking, and the courage to innovate. What good is education if it equips students for exams — but not for life itself?

The Difference Between What's Useful and What's Just Tradition

Let's be honest: not everything traditional should be discarded. There's still a time and place for memorisation. Learning times tables, for example, builds mental agility that supports real-world problem-solving. Foundational literacy and numeracy still matter. So does the ability to write clearly, read with comprehension, and — in today's world — type efficiently. Touch typing isn't optional anymore. It's a core skill in a digital-first economy.

---

[34] **World Economic Forum**, *The Future of Jobs Report* (2023).

But there's a difference between teaching essential tools and clinging to academic traditions that no longer serve students. Algebra, for instance, is still taught as a rite of passage — yet few adults ever use it. We spend years drilling students on the periodic table but rarely connect chemistry to anything beyond exam performance. Shakespeare is studied in depth, yet most students leave school unable to write a persuasive email structure a compelling argument, or understand a contract.

Meanwhile, real-world skills — like budgeting, understanding interest rates, reading pay slips, calculating discounts, comparing insurance policies, or interpreting basic data — are underemphasised or missing entirely. These are the things students will actually use: estimating renovations, decoding workplace jargon, navigating adult responsibilities, and communicating with impact.

Education doesn't need to abandon its past — but it does need to re-examine it through the lens of real-world relevance. Not every lesson should survive just because it's always been there. The question is no longer just about tradition or academic rigour — it's about utility, adaptability, and preparing students to navigate a world that's changed beyond recognition.

What We Can Do Right Now: Practical Steps to Prepare Our Kids

- **Encourage authentic curiosity:** prioritise big questions and bold exploration over correct answers.
- **Develop tech literacy thoughtfully:** foster ethical awareness and creative expression alongside coding skills.
- **Prioritise emotional intelligence:** embed

empathy, collaboration, and self-regulation as core curricula.
- **Champion real-world experiences:** facilitate internships, community projects, and meaningful challenges.
- **Foster entrepreneurial spirit:** inspire innovative thinking, calculated risks, and proactive problem-solving.
- **Support lifelong learning:** celebrate adaptability and resilience as crucial life skills.

It's not just about doing more — it's about daring to do differently. Let's raise children who don't merely fit into the future but actively shape it.

Careers That Sound Like Science Fiction—But Are Already Hiring

Recent reports by Gartner correctly predicted that by 2025,[35] roles involving human-AI collaboration will triple, reflecting the rapid emergence of new, previously unimagined career paths. Job postings demanding AI-related skills surged by 38% between 2020 and 2024, spanning fields from healthcare and finance to retail and education, this isn't a distant forecast — it's unfolding right now. We often ask our children to prepare for a future without clearly depicting its reality. Yet the future is unfolding vividly before us:

- Human-AI Interaction Designers craft emotional connections between humans and technology.
- Digital Wellbeing Coaches foster balanced digital lives amidst pervasive screens.
- Sustainability Engineers design sustainable cities

---

[35] **Gartner**, *Emerging Tech: AI and the Future of Work Trends* (2024).

- and renewable energy solutions.
- Quantum Software Architects pioneer algorithms reshaping computing boundaries.
- Care Economy Innovators reinvent eldercare, childcare, and community systems.
- Global Education Connectors create virtual classrooms that bridge global divides.
- Space Law Strategists draft ethical frameworks for life beyond Earth.

These careers flourish not through memorised answers, but through imaginative questions, compassionate thinking, and brave innovation. This transformation isn't on the horizon — it's already here.

What We're Really Preparing Them For

We aren't simply equipping children for future careers — we're preparing them to lead, innovate, and humanise the future itself. The essential skills aren't technical; they're fundamentally human: creativity, emotional intelligence, adaptability, and deep purpose.

Success won't be claimed by the smartest child in the room, but by the most curious, emotionally intuitive, and courageous. The future belongs to those bold enough to imagine new possibilities and flexible enough to navigate rapid change.

Ultimately, the most important job your child will ever do hasn't even been imagined yet. With the right education, they won't just step into the future — they'll be the ones shaping it. Let's give them the education they truly deserve — the tools to imagine boldly and lead courageously.

Chapter 10: From Action to Vision — Rebuilding Education With Purpose

The Turning Point

The time for observation has passed. The era of patchwork fixes is over. This is the moment we stop tweaking a broken model — and start designing something worthy of the children we claim to care about. This is not just a call to action. It is a blueprint for transformation — rooted in what we know, driven by what we imagine, and powered by what we dare to do. We are not here to renovate the ruins of an outdated system. We are here to rebuild.

Part I: What We Can Do — Today

Transformation doesn't begin with permission. It begins with participation.

Parents: Architects of the Everyday

The first classroom is the home. The first curriculum is conversation. The first test is how we show up. Parents are not spectators — they are stakeholders.

Use your voice. Join school boards. Challenge outdated curricula. Call for education that prioritises inquiry over obedience, depth over coverage, and human development over rigid content. Make learning visible. Learning doesn't begin or end at a school gate. Show children that knowledge lives in gardens, homes, and conversations. Teach them how the world is changing — not just socially or environmentally, but technologically and economically. Let them see the future taking shape — and show them how to navigate, question, and shape it with wisdom.

Give them real-world life skills: how to question what they see online, how to manage money, how to use AI tools ethically, how to adapt to change, and how to think critically in a world flooded with noise. Let curiosity be the compass — but make the map include the landscape ahead, not just the path behind.

Model the mindset. Children become what we model, not what we mandate. Read widely. Reflect openly. Share your failures. Celebrate curiosity. Be the proof that learning never stops. Honour questions, not just answers. A culture of innovation grows where mistakes are safe and questions are sacred. Make your home a haven for both.

Educators: Designers of Possibility

Teachers don't just deliver lessons — they deliver futures. Reimagine the role: you are not a content vessel. You are a curator of wonder, a cultivator of connection, a builder of belonging.

Centre human skills. Create classrooms where communication, empathy, and self-regulation are as valued as maths and grammar. Teach children to lead with compassion and collaborate with purpose.

Rethink rigour. True rigour isn't found in rote memorisation — it's in grappling with nuance, navigating ambiguity, and wrestling with real-world problems.

Personalise with purpose. No learner is average. Tailor pathways. Honour pace. Give students voice, agency, and ownership.

Assess what matters. Not everything that can be counted counts — and not everything that counts can be counted.

We must expand our definition of success. Instead of reducing learning to numbers and averages, we measure what truly matters: adaptability, empathy, ethical decision-making, problem-solving, and the courage to face uncertainty. These are the skills the future demands — and they cannot be bubbled in on a standardised form.

Assessment should mirror the complexity of the world students are entering. Let them demonstrate mastery through what they build, how they reflect, who they collaborate with, and the change they spark. Let feedback be forward-facing. Let growth — not grading — be the standard.

High-trust systems — portfolios, exhibitions, and peer dialogue — become the stage, not standardised snapshots.

Communities: Co-Creators of Culture

We don't educate children in isolation. We do it in ecosystems. Bridge classrooms and communities. Bring local mentors, artisans, scientists, and elders into the learning space. Let students learn from the world — and give back to it. Honour intergenerational wisdom. When children and elders learn side by side, something sacred happens. Stories are shared. Perspectives expand. Connection deepens. Make lifelong learning normal. Create public libraries of skill. Turn cafes into conversation hubs. Host intergenerational workshops. Prove that learning is a lifelong rite, not a childhood phase.

Part II: What We Must Build — Together

The time has come not to reform education, but to redesign it. This next section offers not theory — but vision. Not abstraction — but blueprint. What if learning was guided by purpose instead of compliance? What if schools felt more like studios, labs, gardens, and sanctuaries — and less like factories? What if we dared to design the future, instead of dragging the past behind us?

A Day in a Future Worth Building

The day begins not with a bell — but a question. A 10-year-old logs into a global cohort, collaborating with peers across borders, exchanging real experiences to co-design climate solutions. By midday, they're knee-deep in a river, collecting water samples and testing for quality. By the afternoon, they are presenting their findings to the local council — linking data with story. Science with purpose. This isn't a project. It's participation in the real world. It's agency. It's resonance. It's education reimagined — alive, purposeful, and personal.

Curriculum That Breathes

Subjects are not boxes. They are branches of a living system. The curriculum flows through projects like "The Future of Food" or "Redesigning Democracy," integrating science, humanities, technology, and ethics. Students co-create their learning paths, investigate big questions, and present not just to teachers — but to audiences that matter.

Assessment That Elevates

Grades shrink learners. Evidence expands them. In this future, students build portfolios, lead community exhibitions, reflect on growth, and collaborate on feedback. Assessment becomes a dialogue, not a judgement. It celebrates progress. It illuminates potential. The new question is not "What score did you get?" but "What did this experience teach you — about the world, about others, and about yourself?"

Teachers as Trusted Professionals

Teachers are not data managers — they are design thinkers. They're equipped, resourced, and trusted to lead. They collaborate with community leaders, industry experts, and fellow educators to design truly transformative experiences. Teaching becomes a calling again — not a survival sport.

Technology That Extends Humanity

Tech is not the answer. But it can help us ask better questions. AI becomes an invisible scaffolding — adapting support, unlocking access, mapping growth. VR and AR turn learning into a multisensory experience. And students don't just consume — they create, code, question, and contribute ethically in the digital world.

Wellbeing as Infrastructure

Wellbeing isn't a box to tick. It's the soil everything grows from. Emotional check-ins, mindfulness, play, and circle time aren't bonuses — they're baselines. Nutrition, safety, connection, and belonging are the pulse of every day. Children can't learn if they don't feel safe. They won't create if they don't feel seen.

Global Classrooms, Local Roots

Learning is both wide and deep. Students form global teams to tackle shared challenges. They also learn the stories of their own land, culture, elders, and heritage. They see themselves not just as students — but as stewards. Of knowledge. Of place. Of each other.

Equity, Trust, and Shared Stewardship

A future-ready system doesn't treat equity as charity. It embeds it as structure. Support isn't reactive. It's foundational. Fairness isn't sameness — it's responsiveness. And accountability? It's not top-down. It's a shared agreement between students, teachers, families, and communities. Everyone owns the outcomes.

What It All Adds Up To

A transformed education system doesn't just prepare students for the world as it was — it equips them to shape the world as it must become. It gives them the tools to adapt in a landscape that is shifting beneath our feet. It teaches them to navigate artificial intelligence with ethics, to collaborate across borders with empathy, and to build lives not around job titles, but around meaning.

It doesn't ask them to memorise history — it empowers them to make it. This kind of education doesn't wait for the future to arrive. It helps our children create it — consciously, courageously, and together. It grows thinkers, healers, makers, leaders. Not by accident — but by design. This is not a fantasy. It is already forming. It's not a distant ideal. It is a decision we can make — together, right now. Let this be the generation that doesn't just fix education — but redefines what it means to raise a human being.

Chapter 11: Final Thoughts – Choosing Courage

"We do not inherit the world from our ancestors; we borrow it from our children."

The journey through these pages began with a confrontation — an honest look at what education is, and what it urgently must become. It has led here, not to an ending but to a beginning. The moment has arrived to step forward — not simply with a vision, but with clarity, courage, and conviction. Education shapes more than minds; it shapes the very essence of who young people believe they can become. It holds the power to either confine potential or set it free. It determines whether children learn merely to navigate a world that already exists — or to imagine, create, and lead the one emerging before them.

We are not preparing children for what's coming. The future is already here — coded, automated, and accelerating faster than most of us can comprehend. AI will shape every part of the world our children inherit — with or without our guidance.

Every child enters school carrying dreams, boundless imagination, and innate curiosity. Education's role is not to tame or contain these qualities, but to fiercely protect and nurture them. Classrooms should be places where curiosity ignites rather than quiets, where play and creativity form the backbone of deep, meaningful growth. Spaces where children leave each day more confident, more resilient, and more inspired than when they arrived.

True equity in education moves far beyond equal access — it demands genuine respect, relevant experiences, and a profound sense of belonging for every learner. It recognises the unique brilliance within each child and commits relentlessly to fostering environments where they can thrive as their authentic selves.

Teachers must be supported not as mere implementers of curriculum, but as visionary leaders — trusted to craft transformative experiences that celebrate and elevate human potential. Technology, harnessed wisely and ethically, should amplify rather than overshadow human connection, imagination, and purpose.

Transforming education is not the work of solitary heroes, but the quiet, courageous acts of many. It starts with teachers who choose vulnerability over conformity, parents who advocate fiercely, and leaders who prioritise integrity over ease. It unfolds in classrooms, homes, and communities willing to challenge norms, to speak truths, and to dream bigger.

Our courage now must match the speed of the revolution we're already inside.

Now, a choice stands before everyone touched by education: accept the status quo, or choose the courage to build something extraordinary. Not a system bound by tradition, but one liberated by possibility. Not schools that measure worth through tests alone, but communities where every child knows they matter, their voice is heard, and their dreams have space to flourish.
Education must be our boldest act of hope — our clearest declaration of what humanity is capable of. This moment demands we decide not only what we stand for — but who

we must become. Because education, at its heart, is not merely about preparing children for the future. It is about empowering them to shape it.

Confidently.
Compassionately.
And without apology.

"***Education is the most powerful weapon which you can use to change the world.***"[36] — Nelson Mandela

The sky's not the limit — it's the starting line.

---

[36] **Nelson Mandela**, speech in South Africa, July 16, 2003.

Acknowledgement

To the love of my life, Tim — thank you. Your unwavering support, sharp mind, and deeply compassionate heart have fuelled this book from the very beginning. You challenge me to be better, to think deeper, and to dream more boldly. Our conversations — so often about the future, about purpose, about possibility — are woven into every word of this book. Your voice is in these pages. Your strength is in my courage. This book would not exist without you.

About the Author

Sarah Kissane is a writer, fearless disruptor, and unapologetic truth-teller. Born in the UK and now based in Australia, her path has rarely followed the script — and that's where much of her strength lies.

A single mother by choice through donor conception, Sarah first stepped into authorship with The Special Two — a raw and deeply personal tribute to her son and their extraordinary beginning. Once a struggling reader herself, she went on to earn a BA Honours in Sociology, worked as a childcare educator, supported children with special needs, and taught English in Japan. For over two decades, she has led in the fast-paced world of tech recruitment.

Across every chapter of her life, Sarah has witnessed the systems that shape us — and the cracks we keep ignoring. From classrooms to corporate boardrooms, she's seen how fast the world is changing — and how slow we are to prepare for it.

She writes not to provoke for the sake of noise, but to wake people up. Her work is rooted in love, sharpened by lived experience, and fuelled by an unshakeable belief in humanity's capacity to evolve — if we dare to look honestly at what's no longer working.

Obsolete isn't just a book. It's a mirror. A reckoning. A call to those who know we're overdue for something better — and are ready to be part of building it.

The AI revolution isn't on the horizon.
It's already redesigning the world beneath our feet.

If we surrender our voice now, we surrender our ability to shape what stays human.

This isn't about resisting progress. It's about refusing to disappear inside it. The systems rising around us were not built to protect meaning, nuance, or soul.

And if we don't lift our voice — with intention, urgency, and truth — we risk becoming fluent in technology, but silent in humanity.

This is not just a warning.
It's a call to remember what must remain.

JOIN THE MOVEMENT
The system won't change.
Unless we do.

#obsoletemovement
sarahkissane.com

www.ingramcontent.com/pod-product-compliance
Lightning Source LLC
LaVergne TN
LVHW051511070426
835507LV00022B/3041